In Memory of

My Mother,

Mrs. Betty Moser

From

Jill L. Bowen

Why Why Why does the Earth spin round?

First published as hardback in 2006 by
Miles Kelly Publishing Ltd, Bardfield Centre,
Great Bardfield, Essex, CM7 4SLCopyright

This 2009 edition published and distributed by:

Mason Crest Publishers Inc.
370 Reed Road, Broomall, Pennsylvania
19008
(866) MCP-BOOK (toll free)
www.masoncrest.com

Why Why Why—
Does the Earth Spin Round?
ISBN 978-1-4222-1588-3
Library of Congress Cataloging-in-Publication data is available

Why Why Why—?
Complete 23 Title Series
ISBN 978-1-4222-1568-5

Printed in the United States of America

Contents

Where did the Earth come from?

The Earth came from a cloud of dust. The dust whizzed around the Sun at speed and and began to stick together to form lumps of rock. The rocks crashed into each other to make the planets. The Earth is one of these planets.

A cloud of dust spun around the sun

Lumps of rock began to form

Why does the Moon look lumpy?

Big rocks from space, called meteorites, have crashed into the Moon and made dents on its surface. These dents are called craters and they give the Moon a lumpy appearance.

Look

At night, use some binoculars to look at the Moon. Can you see craters on its surface?

The Earth was formed from the lumps of rock

What is the Earth made of?

The Earth is a huge ball-shaped lump of rock. Most of the Earth's surface is covered by water—this makes the seas and oceans. Rock that is not covered by water makes the land.

Face the Moon!

The Moon travels around the Earth. As the Moon doesn't spin, we only ever see one side of its surface.

Why does the Earth spin round?

Morning

The Earth is always spinning. This is because it was made from a spinning cloud of gas and dust. As it spins, the Earth leans a little to one side. It takes the Earth 24 hours to spin around once. This period of time is called a day.

Evening

Discover

There are 24 hours in a day. How many minutes are there in one hour?

Spinning Earth

Hot and cold!

In the Carribean, the sea can be as warm as a bath. In the Arctic, it is so cold, that the sea freezes over.

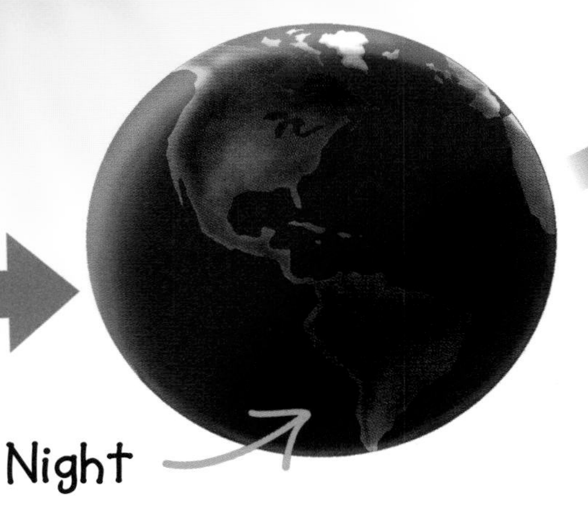

Night

Why do we have day and night?

Every day, each part of the Earth spins towards the Sun, and then away from it. When a part of the Earth is facing the Sun, it is daytime there. When that part is facing away from Earth, it is night time.

Do people live on the Moon?

No they don't. There is no air on the Moon so people cannot live there. Astronauts have visited the Moon in space rockets. They wear special equipment to help them breathe.

What is inside the Earth?

There are different layers inside the Earth. There is a thin, rocky crust, a solid area called the mantle, and a center called the core. The outer part of the core is made of hot, liquid metal. The inner core is made of solid metal.

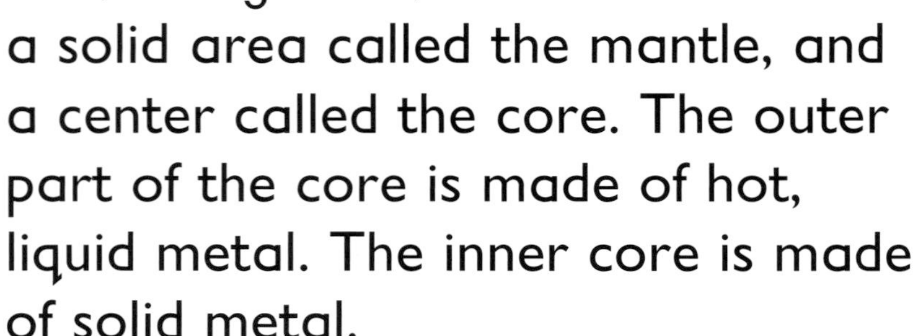

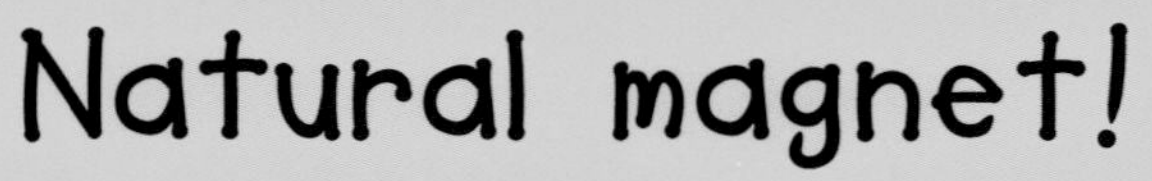

Natural magnet!

Near the center of the Earth is hot, liquid iron. As the Earth spins, the iron behaves like a magnet. This is why a compass needle points to North and South.

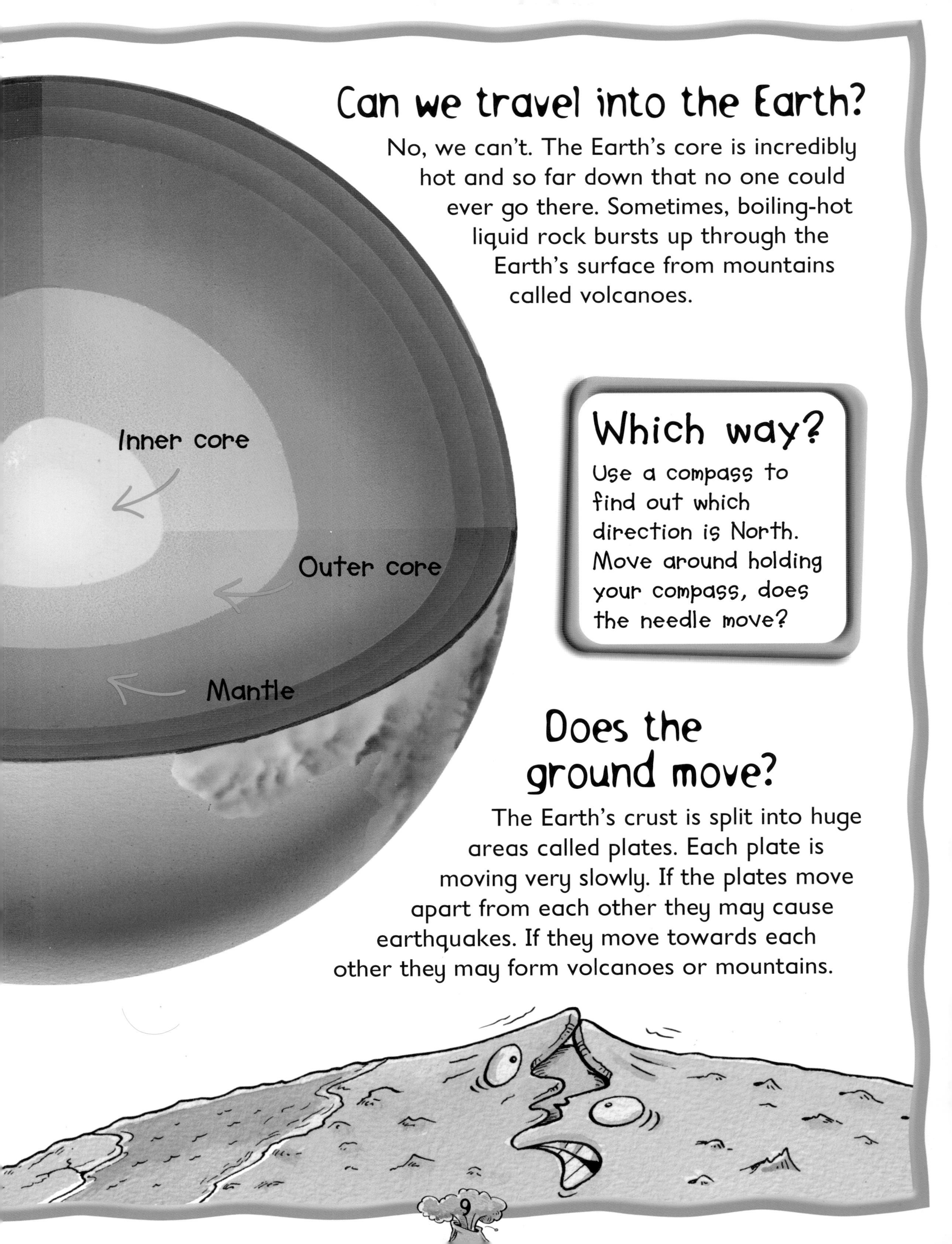

Can we travel into the Earth?

No, we can't. The Earth's core is incredibly hot and so far down that no one could ever go there. Sometimes, boiling-hot liquid rock bursts up through the Earth's surface from mountains called volcanoes.

Which way?

Use a compass to find out which direction is North. Move around holding your compass, does the needle move?

Does the ground move?

The Earth's crust is split into huge areas called plates. Each plate is moving very slowly. If the plates move apart from each other they may cause earthquakes. If they move towards each other they may form volcanoes or mountains.

What is a fossil?

A trilobite was an ancient sea creature

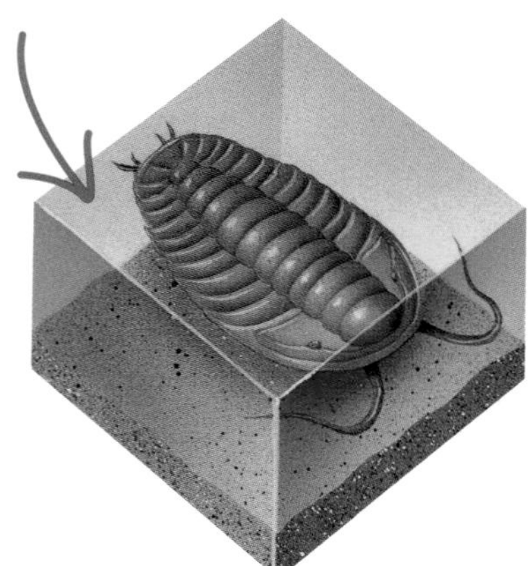

A fossil was once a living thing that has now turned to stone. By studying fossils, scientists can learn more about the past and how animals, such as dinosaurs, used to live.

Scientists digging up and studying fossils

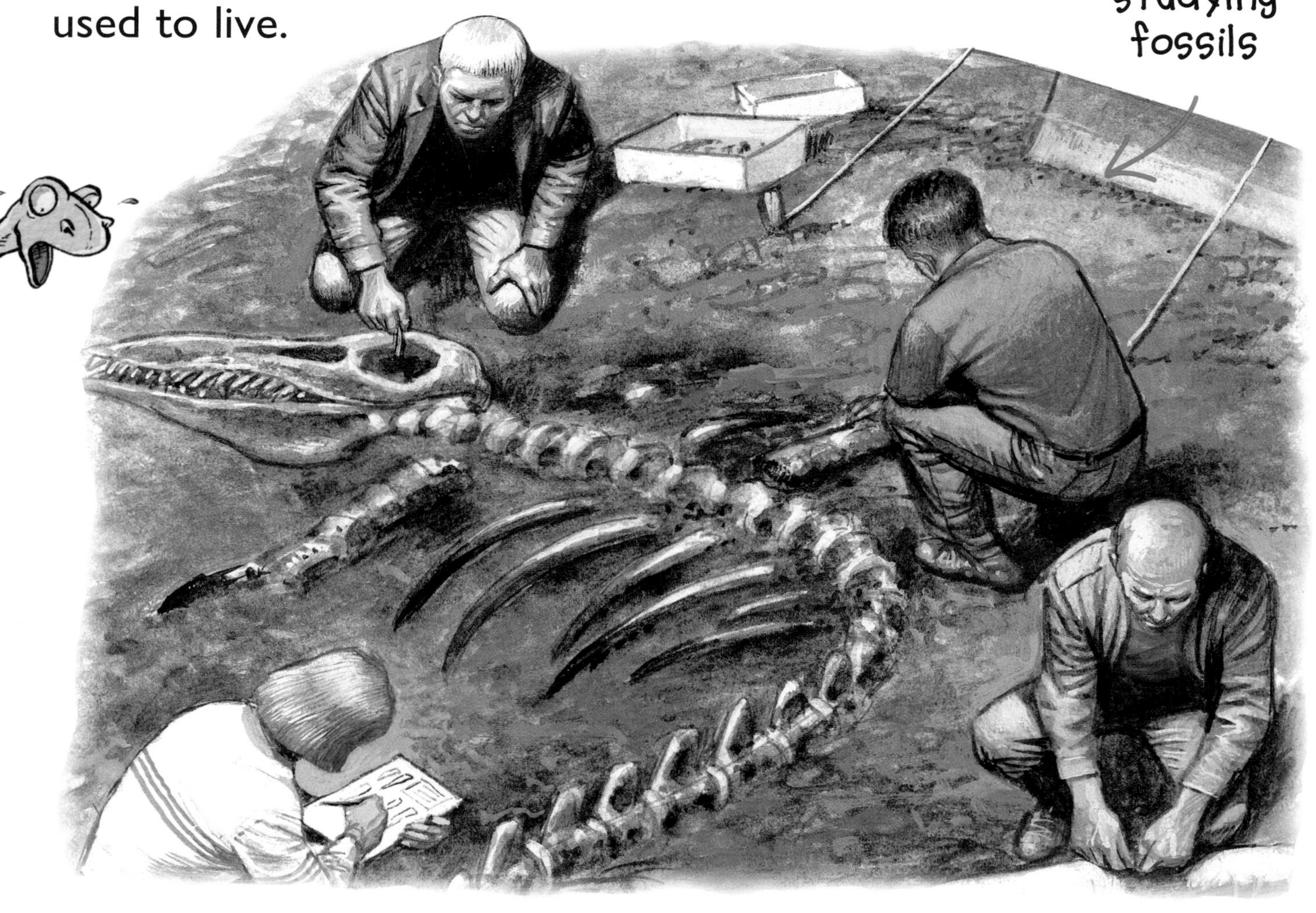

How is a fossil made?

It takes millions of years to make a fossil. When an animal dies, it may be buried by sand. The soft parts of its body rot away, leaving just bones, teeth, or shells. These slowly turn to rock and a fossil has formed.

Find

Look for rocks in your garden. They may be so old dinosaurs could have trodden on them.

1. The trilobite dies

2. The trilobite gets covered with mud

3. The mud turns to stone

4. The fossil forms inside the stone

House of stones!

In Turkey, some people live in caves. These huge cone-shaped rocks stay very cool in the hot weather.

Why do rocks crumble?

When a rock is warmed up by the Sun it gets a little bigger. When it cools down, the rock shrinks to its original size. If this process happens to a rock too often, it starts to crumble away.

What is a volcano?

A volcano is a mountain that sometimes shoots hot, liquid rock out of its top. Deep inside a volcano is an area called a magma chamber. This is filled with liquid rock. If pressure builds up in the chamber the volcano may explode, and liquid rock will shoot out of the top.

Liquid rock

Erupting volcano

Magma chamber

Color

Draw a picture of a volcano erupting. Remember to color the lava bright red.

What is a range?

A range is the name for a group of mountains. The biggest ranges are the Alps in Europe, the Andes in South America, the Rockies in North America, and the highest of all—the Himalayas in Asia.

How are mountains made?

One way that mountains are formed is when the Earth's plates crash together. The crust at the edge of the plates slowly crumples and folds. Over millions of years this pushes up mountains. The Himalayan Mountains in Asia were made this way.

Layer on layer

When a volcano erupts, the hot lava cools and forms a rocky layer. With each new eruption, another layer is added and the volcano gets bigger.

Why are there earthquakes?

Earthquakes happen when the plates in the Earth's crust move apart suddenly, or rub together. They start deep underground in an area called the focus. The land above the focus is shaken violently. The worst part of the earthquake happens above the focus, in an area called the epicenter.

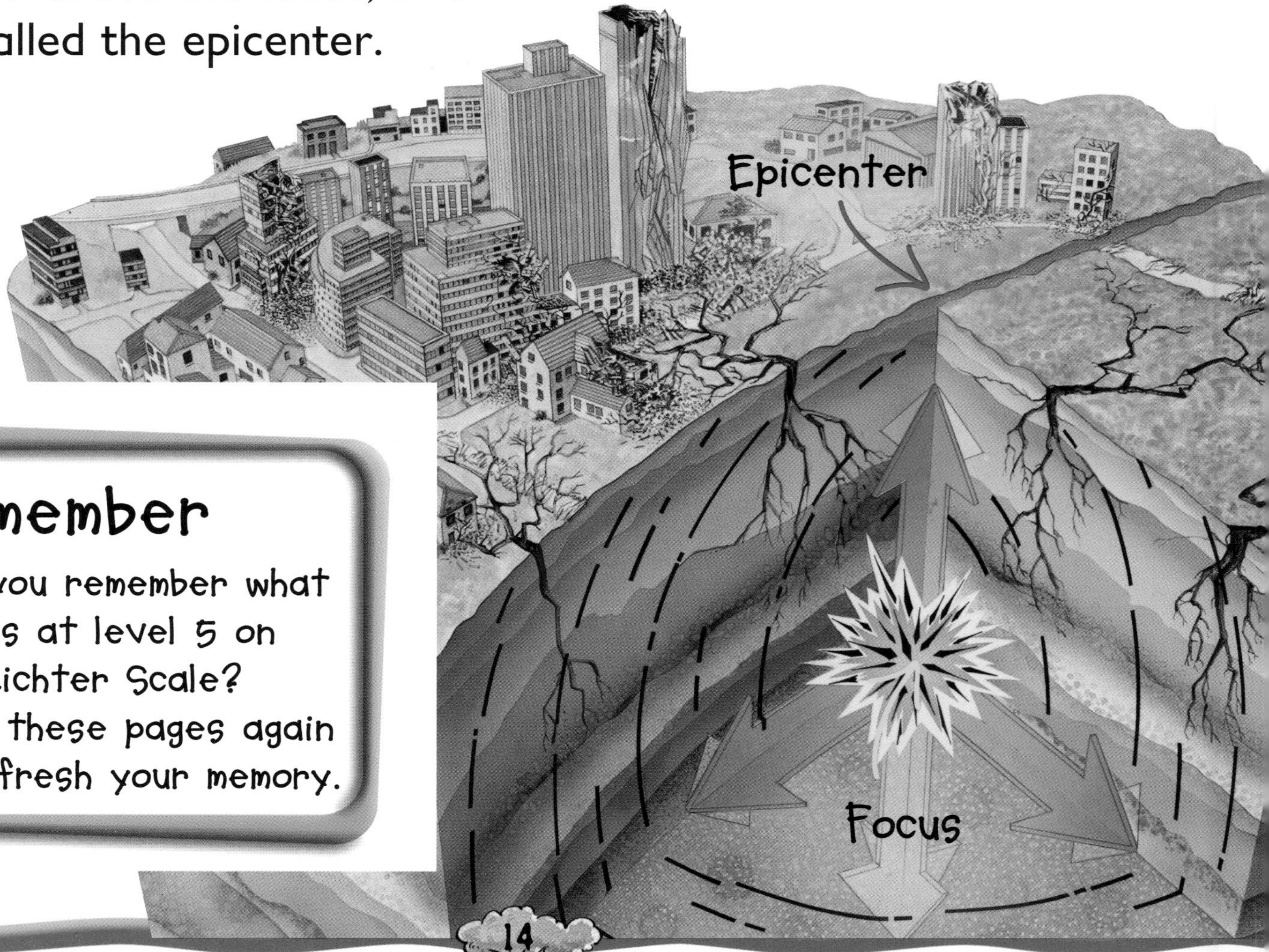

Epicenter

Focus

Remember

Can you remember what breaks at level 5 on the Richter Scale? Read these pages again to refresh your memory.

Lights swing at level 3

Windows break at level 5

Bridges and buildings collapse at level 7

What is The Richter Scale?

The Richter Scale measures the strength of an earthquake. It starts at number one and goes up to number eight. The higher the number, the more powerful and destructive the earthquake.

Super senses!

Some people believe that animals can sense when an earthquake is about to happen!

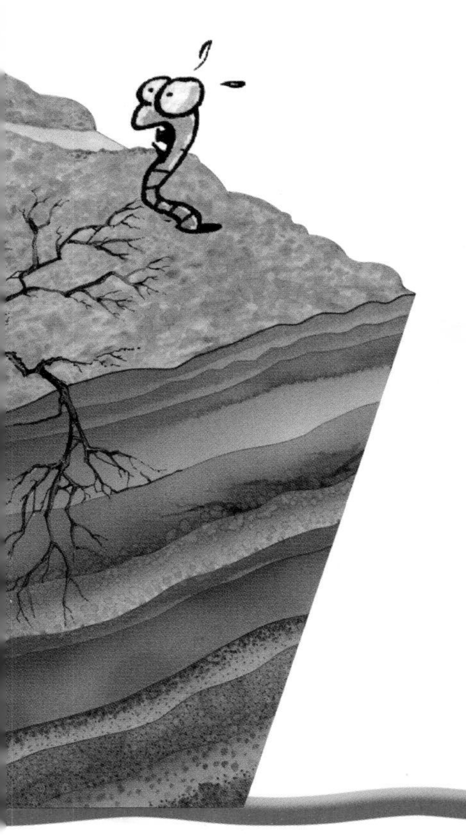

Can earthquakes start fires?

Yes, a powerful earthquake can cause fires. In 1906, a huge earthquake in San Francisco, USA, caused lots of fires. The fires burnt down most of the city and the people who lived there became homeless.

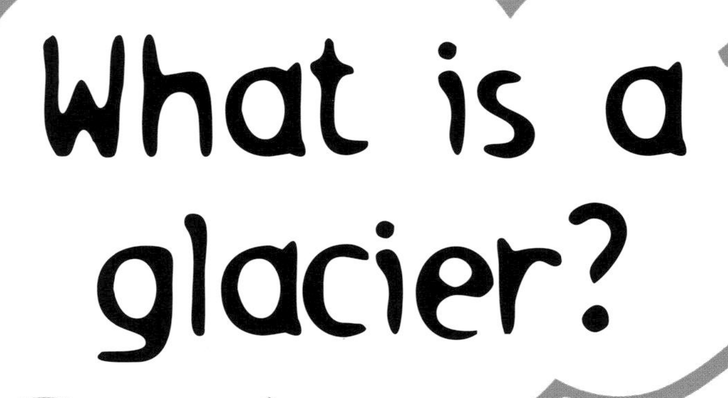

What is a glacier?

Glaciers are huge rivers of ice found near the tops of mountains. Snow falls on the mountain and becomes squashed to make ice. The ice forms a glacier that slowly moves down the mountainside until it melts.

Moving glacier

Fancy flakes!

Snowflakes are made of millions of tiny crystals. No two snowflakes are the same because the crystals make millions of different shapes.

Melted ice

Can ice be fun?

Yes, it can! Many people go ice skating and they wear special boots with blades on them called ice skates. Figure skaters are skilled athletes who compete to win prizes.

What is an iceberg?

Icebergs are big chunks of ice that have broken off glaciers and drifted into the sea. Only a small part of the iceberg can be seen above the water. The main part of the iceberg is hidden under the water.

Look

Next time it snows, put some gloves on and let the snowflakes fall into your hand. Can you see crystals?

Where do rivers flow to?

Rivers flow to the sea or into lakes. They start off as small streams in hills and mountains. The streams flow downhill, getting bigger and wider. The place where a river meets the sea, or flows into a lake, is called the river mouth.

Oxbow lake

River mouth

Why are there waterfalls?

Waterfalls are made when water wears down rocks to make a cliff face. The water then falls over the edge into a deep pool called a plunge pool. Waterfalls may only be a few inches high, or several hundred feet high!

Discover

Try to find out the name of the highest waterfall in the world. Where is it?

Risky business!

Salmon are a type of fish. Every year, fishermen try to catch them as they swim back to the river they were born in to have their babies.

What is a lake?

A lake is a big area of water that is surrounded by land. Some lakes are so big that they are called inland seas. Most lake water is fresh rather than salty. The biggest lake in the world is the Caspian Sea in Asia.

Are there mountains under the sea?

Ocean

Yes there are. Mountains lie hidden in very deep oceans. The ocean floor is very flat and is called a plain. Large mountain ranges may rise across the plain. Some oceans even have underwater volcanoes.

Exploring underwater!

Scientists can learn more about life underwater by exploring the ocean in submarines. They can be underwater for months at a time.

Why do coasts change?

The coast is where the land meets the sea, and it is always changing. In many places, waves crash onto land and rocks, slowly breaking them up. This can change the shape of the coastline.

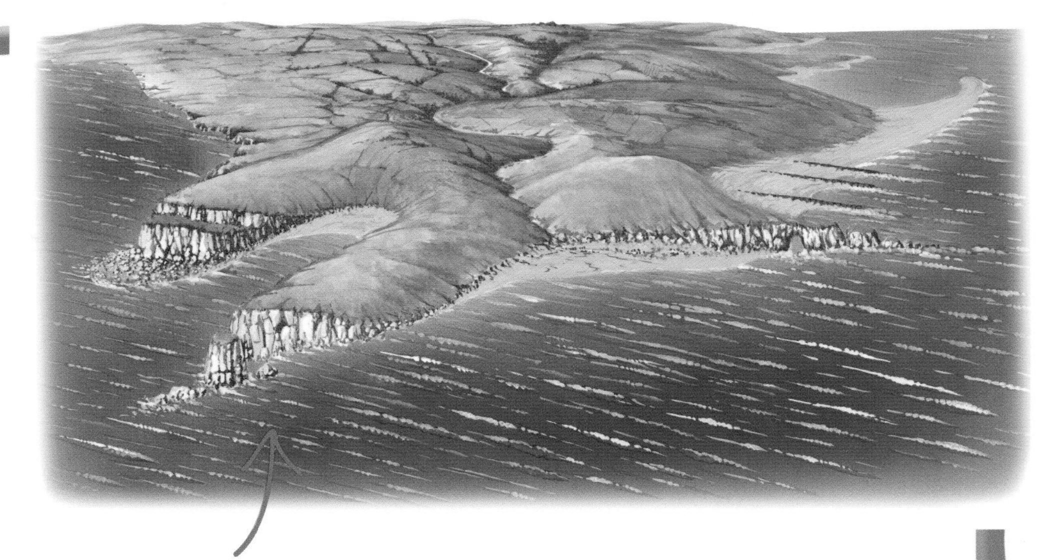

Coastline

Find out

Have a look in an atlas to find out which ocean you live closest to.

Plain

Underwater volcano

Trench

What is coral?

Coral is made from polyps. These are tiny creatures the size of pin heads that live in warm, shallow waters. The polyps join together in large groups and create rocky homes. These are called coral reefs.

How are caves made?

When rain falls on rock, it can make caves. Rainwater mixes with a gas in the air called carbon dioxide. This makes a strong acid. This acid can attack the rock and make it disappear. Underground, the rainwater makes caves in which streams and lakes can be found.

Remember

Stalactites hold on tight, stalagmites might reach the top!

Can lava make caves?

When a volcano erupts and lava flows through the mountain, it can carve out a cave. A long time after the eruption, when the volcano is no longer active, people can walk through this lava cave without having to bend down.

What is a stalactite?

Rocky spikes that hang from cave ceilings are called stalactites. When water drips from the cave ceiling, it leaves tiny amounts of a rocky substance behind. Very slowly, over a long period of time, this grows into a stalactite.

Super spiky!

Stalagmites grow up from the cave floor. Dripping water leaves a rocky substance that grows into a rocky spike.

Is there water in the desert?

Yes there is. Deserts sometimes get rain. This rainwater seeps into the sand and collects in rock. The water then builds up and forms a pool called an oasis. Plants grow around the oasis and animals visit the pool to drink.

What are grasslands?

Grasslands are found when there is too much rain for a desert but not enough rain for a forest. Large numbers of animals can be found living and feeding on grasslands, including zebras, antelopes, and lions.

Create a picture of a camel crossing a desert—don't forget to include its wide feet!

Big feet!

Camels have wide feet that stop them sinking into the sand. They can also store water in their bodies for a long time.

Rainforest

What is a rainforest?

In hot places, such as South America, grow areas of thick, green forest. These are rainforests, and they are home to many amazing plants and animals. Rainforests have rainy weather all year round.

Where does rain come from?

Rain comes from the ocean! Water moves between the ocean, air, and land in a water cycle. A fine mist of water rises into the air from the ocean and from plants. This fine mist then forms clouds. Water can fall from the clouds as rain.

Water falls from clouds as rain

Water cycle

Stormy weather!

Every day there are more than 45,000 thunderstorms on the Earth! Thunderstorms are most common in tropical places, such as Indonesia.

Pretend

Pretend to be a tornado —and watch people get out of your way very quickly!

Tornado

How does a tornado start?

A tornado is the fastest wind on Earth. Tornadoes start over very hot ground. Here, warm air rises quickly and makes a spinning funnel. This funnel acts like a vacuum cleaner, destroying buildings and lifting cars and trucks off the ground.

A fine mist of water rises from the ocean

Do storms have eyes?

Yes, storms do have eyes! A hurricane is a very dangerous storm. The center of a hurricane is called the eye and here it is completely still. However, the rest of the storm can reach speeds of up to 190 miles an hour.

How do we damage the Earth?

Some of the things that people do can damage the Earth. Factories pump chemicals into the air and water. Forests are being cut down, killing the wildlife that lives there, and fumes from cars are clogging up the air. Scientists are trying to find new ways to protect the Earth before it is too late.

Fumes from factories

Help

Save all of your empty drinks cans and bottles and take them to your recycling center.

Waste dumped in rivers

Traffic fumes

Save the planet!

Large areas of land have been made into national parks where wildlife is protected. People can go there to learn about both plants and animals.

1. Old bottles are collected from bottle banks

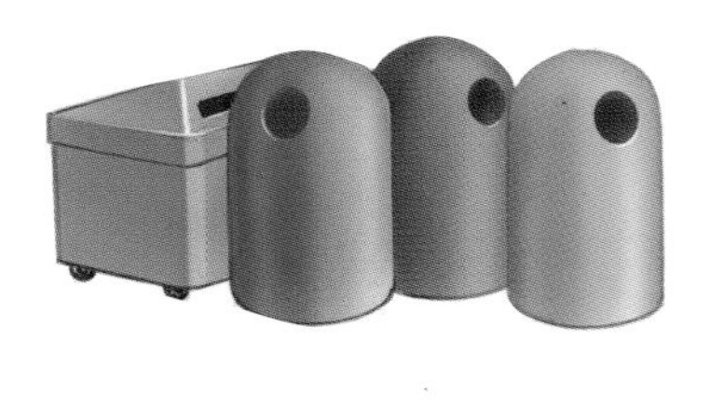

What is recycling?

Recycling is when people collect waste materials, such as paper and plastic. The waste is then taken to a recycling center and changed back into useful materials to make many of the things we use today.

2. The glass or plastic are recycled to make raw materials

3. The raw materials are reused to make new bottles

Trees are cut down

How can we protect the Earth?

There are lots of things we can do to protect our planet. Recycling, picking up litter, switching lights off, and walking to the shops all help to make a difference.

Quiz time

Do you remember what you have read about planet Earth? These questions will test your memory. The pictures will help you. If you get stuck, read the pages again.

page 4

1. Why does the Moon look lumpy?

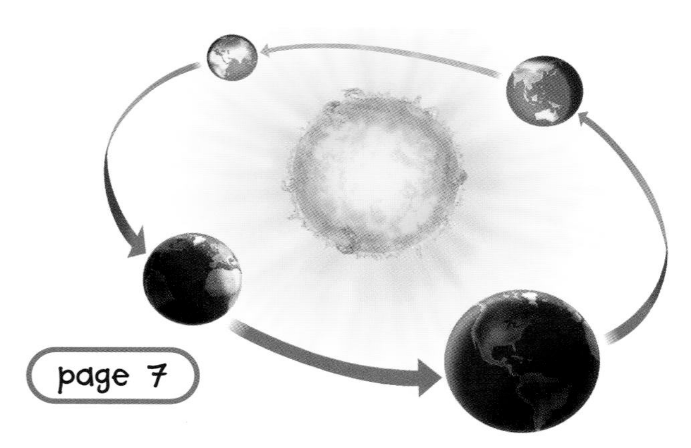

page 7

2. Why do we have day and night?

3. What is inside the Earth?

page 8

4. What is a fossil?

page 10

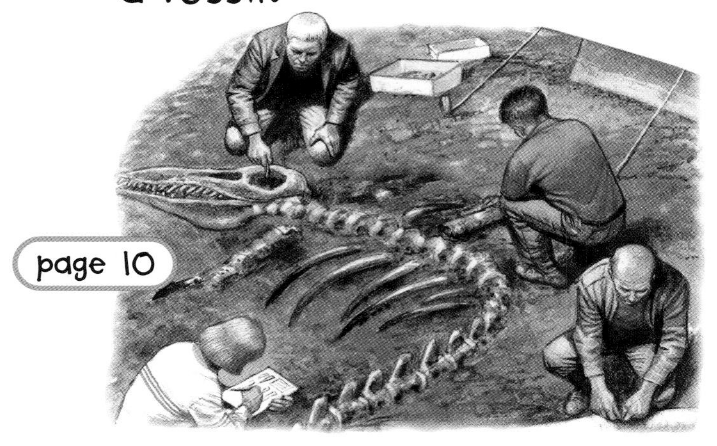

5. What is a range?

page 13

6. Can earthquakes start fires?

page 15

7. Can ice be fun?

page 17

page 17

8. What is an iceberg?

page 18

9. Why are there waterfalls?

page 23

10. Can lava make caves?

page 23

11. What is a stalactite?

page 25

12. What is a rainforest?

page 27

13. Do storms have eyes?

Answers

1. Because there are craters on its surface
2. Because the Earth is always spinning
3. Lots of different layers of rocks
4. A living thing that has turned to stone
5. A group of mountains
6. Yes, they can
7. Yes, people skate on ice for fun
8. A big chunk of ice that has broken off a glacier
9. Because water flows over cliffs
10. Yes, it can
11. A rocky spike that hangs from the ceiling of a cave
12. A large, thick, green forest that grows in a hot place
13. Yes, the center of a hurricane is called the eye

Index